I0070690

James M. Anthony

13 Tips to be more Cybersecure

Copyright © 2024 by James M. Anthony

All rights reserved. No part of this publication may be reproduced, stored in a retrieval system, or transmitted, in any form or by any means, electronic, mechanical, photocopying, recording or otherwise, without the prior permission of the author.

ISBN 979-8-9916938-2-0

"If you aren't concerned about Cybersecurity, you don't know enough about it."

Raef Meeuwisse, Cybersecurity & Influencing Speaker

CONTENTS

FOREWORD

2016 was the year when the number of mobile phone subscriptions overtook population figures. Ever since, there have been more smartphones than people on our planet – 8.58 billion vs. 7.95 billion in 2022 [World Economic Forum, 2023]. Furthermore, around two-thirds of the global population is currently connected to the World Wide Web: as of July 2024, there were 5.45 billion Internet users and 5.17 billion social media users worldwide [Petrosyan, 2024].

Out of those 5+ billion people using 8+ billion smartphones (aside from laptops, tablets, smartwatches, and a plethora of additional electronic devices we commonly use in our everyday life), how many are fully aware and mindful of the cyber threats they are facing?

I bet only a relatively small fraction of them.

Be it for innocence, inexperience, gullibility, ignorance, or thoughtlessness, the number of victims of cybercrime is constantly on the rise. According to Forbes, there were 2,365 cyberattacks in 2023, with over 340 million victims [St. John, 2024].

This book targets all those vulnerable people out

there, who could be tomorrow's victims of cybercriminals. I will share with you 13 tips that will make you more aware, and your digital tools and data more secure.

I will not use technical verbiage or jargon. This book is not meant for cybersecurity experts nor for organizations. A vast literature exists that discusses how companies should develop and implement their cybersecurity strategies. This book is for all the John Does in the world, who tap their touchscreens or type on their keyboards without even knowing of how to spell the word "cybersecurity".

There is no such thing as being 100% secure. Even the most experienced cybersecurity professional can be fooled, and the most well-designed system can be breached. However, we can make our devices and systems harder to access and more resilient to cyberattacks.

This book is a short and easy read. It should not take you more than two or three hours to go over it. Nonetheless, if any of my tips will prevent even one of you from falling victim to cybercrime, I will have achieved my purpose.

Enjoy the reading!

James

CYBERSECURITY

When ARPANET, the precursor of Internet, moved its first steps in the late 1960s and early 1970s, it was little more than a small network of researchers, with no compelling need to secure access or protect data. It was not until 1988 that the first significant cyberattack occurred, with the *Morris Worm* infecting around 6,000 computers, about 10% of all Internet-connected computers at the time [Cristello, 2023]. The Morris Worm was the first malicious software (or *malware*) in history.

As the Internet grew, so did the potential for unauthorized access and data breaches, with the disclosure of sensitive information. This led to the implementation of cybersecurity legislation and regulations, and the development of new security measures, such as anti-virus software, password protection, data encryption, as well as systems to detect and prevent cyber threats in real time. But it was only after the year 2000 that many cybersecurity companies like *PaloAlto Networks* and *CrowdStrike* emerged, providing effecting solutions to combat cybercrime.

The first hackers were often young, nerdish students working out of their basements or garages, as in the case of David Lightman, the leading character in the 1983 movie *WarGames*.

The scenario changed dramatically over the following decades, with a shift from individual cybercrime to organized cyberattacks, often funded or orchestrated by **Nation States** as a form of warfare. Famous examples are the 2007 Stuxnet attack that destroyed numerous Iranian nuclear centrifuges, the 2015 Ukraine's power grid hack that caused a power outage for more than 230,000 people, and the 2024 coordinated explosions of thousands of pagers and walkie-talkies used by Hezbollah operatives in Lebanon.

The rise of cyber warfare has demonstrated the potential for cyberattacks not only to disrupt the digital world, but also to create significant damage, death, and destruction in the physical world.

Nation states are only one of the many threat actors in today's cybersecurity landscape. Names and categories vary slightly depending on the sources, but the main ones are [IBM, 2024]:

- **Cybercriminals**: individuals or organized groups acting mostly for financial gain.

- **Cyberterrorists**: usually politically or ideologically motivated, they carry out attacks that threaten or result in violence.

- **Hacktivists**: they aim at promoting political or social agendas, like freedom of speech (you probably heard of *Anonymous*, haven't you?). The name is an amalgamation of the words *hacker* and *activist*.

- **Insider Threats**: often disgruntled employees who act for monetary gain or retaliate against their employer.

- **Thrill Seekers**: they use hacking techniques primarily for fun or personal satisfaction.

Such individuals or groups can cause significant financial and/or reputational damage to both private citizens and organizations, by stealing sensitive data, or encrypting it in exchange for a ransom, disrupting services or activities, damaging hardware, spying on their victims.

Today, in a world where most people have (at least) an email account and a smartphone but often poor cybersecurity awareness, a world where *smart* devices connected to the Internet are everywhere in our homes and vehicles, threat actors find fertile ground for their scams, that become increasingly more sophisticated and difficult to detect.

We cannot predict with certainty what the future holds for us, but new technologies such as quantum computing, artificial intelligence, and 5G – the fifth generation of wireless technology – are likely to play a crucial role in cybersecurity, since they can be used on both sides of the battlefield, by threat actors as well as by security professionals, to either develop new exploits or to enhance cybersecurity defenses.

In this rapidly evolving landscape, raising public awareness about cyber threats and safe online practices is paramount.

This is what we will focus on in the next 13 chapters.

Tip #1:

Use strong, unique passwords

According to NordPass, an average person like you or me has around 100 passwords [Silkalns, 2024].

One. Hundred.

As inflated this number may sound, think about how many electronic devices we use in our everyday lives: smartphones, desktops and laptops, tablets, gaming consoles, smartwatches and fitness trackers, smart TVs, digital cameras, Wi-Fi routers, smart home devices. And the list could go on and on.

Also, consider how much we rely on websites and apps for communicating with family and friends, posting on social media, booking flights and hotels, ordering food, watching movies, listening to music, getting travel or food advice, checking our bank accounts, calling a taxi, purchasing tickets to events, learning languages, reading newspapers and books, finding a partner, applying for jobs, and lots of other tasks. Most websites and apps require users to create an account or log in.

With such a humongous quantity of electronic

devices and accounts, **passwords** are your first line of defense to protect your personal data against cybercriminals.

But passwords can be *guessed*, they can be *leaked*, they can be *forgotten*. The respective solutions to these 3 problems are strong passwords, unique passwords, and password managers. Let's take a look at each one of them.

Strong Passwords

A cybercriminal could guess your password by simply trying any possible combination of characters until they eventually find the right one. This trial-and-error method is called *Brute-Force Attack*. Brute-Force Attacks are usually carried out by *botnets* (short for *robot networks*), groups of hijacked computer devices (*bots*) connected to the internet that are under the control of a single attacking party, known as the *bot-herder*. Botnets can make thousands of password guesses per second.

Depending on how strong it is, your password may require thousands of years to get cracked, or just a few seconds.

But what does *strong* mean when referred to a password?

Experts agree that password **length** is the primary factor characterizing its strength: the longer the password, the harder it is to crack. NIST (the United States National Institute for Standards and Technology) sets the minimum length to 8 characters [NIST, 2017]. Microsoft recommends at least 12 characters [Microsoft, 2024].

2 Uppercase	2 Lowercase	2 Digits	2 Symbols	8 Characters

1aH$5g/M

Enter and edit your test passwords in the field above while viewing the analysis below.

Brute Force Search Space Analysis:

Search Space Depth (Alphabet):	26+26+10+33 = **95**
Search Space Length (Characters):	8 characters
Exact Search Space Size (Count): (count of all possible passwords with this alphabet size and up to this password's length)	6,704,780,954,517,120
Search Space Size (as a power of 10):	6.70×10^{15}

Time Required to Exhaustively Search this Password's Space:

Online Attack Scenario: (Assuming one thousand guesses per second)	2.13 thousand centuries
Offline Fast Attack Scenario: (Assuming one hundred billion guesses per second)	18.62 hours
Massive Cracking Array Scenario: (Assuming one hundred trillion guesses per second)	1.12 minutes

Note that typical attacks will be online password guessing limited to, at most, a few hundred guesses per second.

GRC's Interactive Brute-Force Password "Search Space" Calculator [GRC, 2012].

An 8-character password consisting of upper- and lowercase letters, numbers, and symbols has 95 options[1] for each character. This means that hackers carrying out a Brute-Force Attack will be challenged with more than 6.7 quadrillion ($6.7*10^{15}$) possible combinations. As shown in the picture above, the Gibson Research Corporation estimates in over 200,000 years the time required to exhaustively search such space in case of an online attack [GRC, 2012].

A couple of recommendations to make your passwords more difficult to guess:

- **The longer, the better**. If the website, app, or device allows, choose passwords longer than the minimum 8 characters recommended by NIST. 12 or more characters are advisable. The UK National Cyber Security Centre (NCSC) suggests combining 3 random words to create passwords (example: *applenemobiro*) that are long enough and strong enough [NCSC, 2021].

- **Long does not mean impossible to remember**. A password like *u/7zFo;W3t* is not stronger than *Aut0?B4gs!*. Both have 10 characters – 4 lowercase

---

[1] 26 lowercase letters (*a* to *z*), 26 uppercase letters (*A* to *Z*), 10 numbers (*0* to *9*), and 33 special characters (*! " # $ % & ' ( ) * + , - . / : ; < = > ? @ [\ ] ^ _ ` { | } ~* plus the *space* character).

letters, 2 uppercase letters, 2 numbers, 2 special characters –, but the second password is definitely easier to remember.

- **Avoid words that are listed in a dictionary or words that are too common**. Hackers execute *Dictionary Attacks*, a special type of Brute-Force Attack that consists in systematically trying any possible word listed in a dictionary as potential password. Picking up a word in a dictionary and using it as a password is therefore not advisable. Also, the word *password* itself has 8 characters, but it is so common that it is an extremely weak password (sadly, it is still the most popular password worldwide, at a count of nearly 5 million [Silkalns, 2024]). The same is true for another popular (but very weak) password: *12345678*.

- **Never use personal information**, such as your birthday or your cat's name as a password, as it can be easily predicted by cybercriminals, especially if they are watching you on social media.

Unique Passwords
Passwords can be stolen, and at that point it does not matter any longer how strong they are: you need to change them as soon as possible.

Password theft can happen due to your negligence or due to a data leak.

- **Negligence.** If you write down your passwords and leave them in open sight, chances are that someone who has access to your space can steal them. Also, if you share your passwords with someone else, it is possible that this person (often a scammer you met online) uses your credentials to harm you, such as accessing your online bank account and stealing your money.

- **Data Leaks.** Most data leaks happen when an organization's account system is compromised. Unfortunately, millions of sensitive records are stolen every day. Just to give a couple of examples, *LinkedIn* and *Yahoo* were victims of data breaches (in 2012 and 2014, respectively) that led to millions of user log-in credentials being stolen [Breachsense, 2024]. You can check if your email is in a data breach on https://haveibeenpwned.com/. Leaked passwords are often on sale on the *Dark Web*, a virtual space where malicious actors transact illegal products and services.

As soon as you realize that your password may have been stolen, or you are notified of a data leak, change it immediately. The problem arises if you use that very same password for more than one account. In this case, you may fall victim to a *Credential Stuffing Attack*.

Credential Stuffing is a type of cyberattack where hackers use automated tools to try stolen username-password combinations across different websites and services to test if log-in credentials have been reused. Once the cybercriminals gain access to an account using valid credentials, they can cause any kind of damage, including financial theft, data exfiltration, and malware deployment. This technique is extremely dangerous because it leverages human behavior, and specifically our tendency to reuse passwords.

> *"Passwords are like underwear: don't let people see it, change it very often, and you shouldn't share it with strangers."*
>
> **Chris Pirillo**, American Entrepreneur

To cope with Credential Stuffing Attacks, you should use unique (strong) passwords for every website, application, and system you use.

Unfortunately, according to a Verizon Data Breach Investigation, even though 91% of people are aware of the risks of reusing passwords across multiple accounts, 59% do so at both their place of employment and home [Trevino, 2022].

Password Managers

Given the number of devices and accounts we deal with in our everyday life, it is virtually impossible to remember all our passwords by heart, especially once we ensure that each of them is strong and unique.

Writing them down on a piece of paper or a notebook is not necessarily a bad choice, as long as the book is kept in a secure place and nobody but you knows where it is. The chance that a cybercriminal accesses your home and robs your password book are minimal. Of course, the risk increases if you carry the notebook around in public. Someone might steal it or look over your shoulder and see the passwords.

Fortunately, there is a better option for managing passwords securely: a **Password Manager**.

A Password Manager is an app on your computer, tablet, or smartphone that automatically generates strong passwords for all your accounts and stores them in encrypted vaults. You only need to memorize the *master password* to access the Manager, which can be a strong, unique password that you create yourself. Many Password Managers can also enter your passwords into websites and apps automatically, so you do not need to type them when you log in [NCSC, 2021].

There are many Password Managers available on the market, some for a price as low as $2-3 per month, some even for free. The New York Times recommends *1Password* among the paid Password Managers, and *Bitwarden* among the free ones [Eddy, 2024].

One final comment before wrapping up this chapter. Many organizations force employees to periodically change their passwords, usually every three months. This is not what NIST and most cybersecurity experts recommend [NIST, 2017]. According to NIST, there is no real benefit to changing passwords frequently. As long as passwords are strong and unique to each service and account you use, you can probably keep them for life. Unless they are compromised in a data breach, then you should change them immediately [Griffith, 2021].

✓ KEY TAKEAWAYS

1. *Password length is strength*
2. *Long doesn't mean impossible to remember*
3. *Don't use dictionary words nor words that are too common*
4. *Never use personal information*
5. *Don't leave your passwords in open sight*
6. *Don't share your passwords with others*
7. *Change your password only if compromised*
8. *Use a unique password for every account/device you have*
9. *Use a Password Manager*

Tip #2:

Two(-Factor) is better than one

As we discussed in the previous chapter, having strong, unique passwords for each of your accounts is a necessary but insufficient condition to grant you (cyber-)peace at night. Unfortunately, even the most complex password, having enough time and computing power, can be brute forced. Additionally, a password can be stolen, either because you incautiously left it in plain sight or shared it with someone else, or because a website's database got breached and user credentials compromised.

On July 4th, 2024, in what has been so far the largest password dump in history, almost 10 billion (yes, *TEN BILLION*) passwords have been leaked by a hacker. The breach, known as *RockYou2024*, compiled into a single, massive text file passwords that have been likely collected from more than 4,000 databases over the past 20 years [exterro, 2024]. Leaked passwords are now circulating in hacker forums on the *Dark Web*, increasing the likelihood of unauthorized access to numerous accounts.

Implementing **Two-Factor Authentication (2FA)** – or, more in general, **Multi-Factor Authentication (MFA)** – adds an additional layer of security beyond a password and a username, and reduces the likelihood of unauthorized access due to compromised credentials.

> *"Turn on all security features like two-factor authentication. People who do that generally don't get hacked. Don't care? You will when you get hacked. Do the same for your email and other social services, too."*
>
> **Robert Scoble**, Blogger, Technical Evangelist, Author

When you sign into one of your online accounts – be it *Facebook*, *Amazon*, *LinkedIn*, or *Twitter*, just to name a few – you are proving to the service that you are who you say you are. This process is called *authentication*. With 2FA, you need a second verification method – a second *factor* – to prove your identity. Think about withdrawing money at an ATM. This is an example of 2FA, since the ATM asks for your debit card and your PIN.

The 3 most common authentication factors are:

- **Knowledge Factor**, or something you know, like a password or a PIN.
- **Possession Factor**, or something you have, like a smartphone, a credit or debit card, or a token device.
- **Inherent Factor**, or something you are, like a fingerprint, facial recognition, or voice verification.

There are at least two additional factors worth mentioning, although not as common as the other three: the **Location Factor** (or somewhere you are) and the **Behavior Factor** (or something you do). The former verifies where you are by viewing your IP address or GPS location. Banks often use geolocation to detect potentially fraudulent purchases made in places that look suspicious. The latter verifies a user's identity based on unique aspects of their behavior, like how they type or move the mouse [Aratek, 2023].

Simply put, 2FA typically demands you to provide something you have or something you are in addition to something you know. MFA demands you to provide 2 or more factors.

Common 2FA/MFA methods include [Ibrahim, 2024]:

- *One-time passwords* (OTP) sent via text message, email, or automated phone call.
- *Authentication apps* (e.g., Google Authenticator) that generate time-based one-time passwords (TOTP).
- *Hardware tokens* (e.g., YubiKey) that send unique codes or use cryptographic techniques.
- *Biometric verification* that uses fingerprints, facial features, voice patterns, or other physical characteristics.
- *Push notifications* to pre-registered devices, which must be approved by the user.

Nowadays, 2FA is widely used by online banking services (e.g., *Bank of America*), payment processors (e.g., *PayPal*), e-commerce platforms (e.g., *Amazon*), email providers (e.g., *Google*), cloud services (e.g., *Dropbox*), social media platforms (e.g., *Facebook*), and many more. Speaking of social media, here below the instruction to turn on 2FA on *Facebook*, *Instagram*, and *LinkedIn*, in case you have not done it already.

Facebook
1. Click the **Account** icon on the top right of the *Facebook* homepage
2. Click ✿ (**Settings & privacy**)

3. Click anywhere inside the **Accounts Center** box
4. Select **Password and security**, then **Two-factor authentication**
5. If you have more than one account, choose the one you'd like to set up two-factor authentication for
6. (You may be requested to enter your password)
7. Choose a security method (**Authentication app** or **Text message**) and follow the instructions to complete set up

Instagram

1. Click your account icon on the bottom right of the *Instagram* homepage, then click ☰ on the top right
2. Click anywhere inside the **Accounts Center** box
3. Select **Password and security**, then **Two-factor authentication**
4. If you have more than one account, choose the one you'd like to set up two-factor authentication for
5. Choose a security method (**Authentication app** or **SMS or WhatsApp**) and follow the instructions to complete set up

LinkedIn

1. Click the ⓪ **Me** icon on the top left of the *LinkedIn* homepage
2. Select **Settings** at the bottom of the menu
3. Click **Sign in & security**, then click **Two-step verification**
4. Click **Set up** to activate two-step verification
5. Select **Authenticator App** or **Phone Number (SMS)** as verification method
6. If you selected **Phone Number (SMS)** in the previous step, add your phone number, then click on **Send code**
7. Enter the verification code sent to your phone, and follow the instructions to complete set up

---

✓ **KEY TAKEAWAYS**

1. Passwords can be brute-forced or compromised

2. Five Factors: something you know, something you have, something you are, somewhere you are, something you do

3. Turn on 2FA wherever possible

Tip #3:

Change default credentials

The Internet of Things

The number of **IoT (Internet of Things) devices** raised to nearly 16 billion worldwide in 2023, and it is forecasted to double to more than 32 billion by 2030 [Vailshery, 2024].

IoT refers to physical objects that collect and exchange data over the Internet. They include smartwatches like Apple's *iWatch*, voice controllers like Amazon's *Alexa*, smart home devices (e.g., locks, security cameras, thermostats, light bulbs, smoke and carbon monoxide detectors, air quality monitors), smart appliances (e.g., TVs, refrigerators, coffee machines, ovens), irrigation systems, robotic vacuum cleaners and lawn mowers, baby monitors, etc. Most likely, you own one or more of them already.

IoT devices bring many benefits, ranging from task automation (e.g., a robot mowing your lawn) to energy efficiency and cost savings (e.g., smart bulbs turning the lights on only when someone is in the room), to a healthier living environment (e.g., smart air purifiers and thermostats controlling air quality, humidity, and temperature in your home).

The flip side of the coin is the cyber risks that IoT devices bring with them.

In 2017, hackers tried to steal data from a casino in a very unconventional way: by compromising a smart sensor regulating temperature, salinity, and feeding schedules in a high-tech fish tank connected to the Internet. Once cybercriminals gained control of the sensor, they moved laterally to other devices in the casino's network [Larson, 2017].

The year before, in 2016, a botnet of Internet-connected smart home devices – infected by a malware named *Mirai* after a Japanese cartoon – took down CNN, Netflix, Twitter, and other popular websites.

In both cases, the attacks were launched through IoT devices. What allowed the Mirai botnet to spread to around 300,000 devices – and maybe hackers to sneak into the casino's network – is the widespread use of **default usernames and passwords** in smart devices, as well as in database systems, Industrial Control Systems (ICS), software packages (e.g., videogames), and network devices like routers, switches, and firewalls [CISA, 2016]. Attempting to log in with default passwords is a widely used attack technique.

Many **printers** come with default settings too. Considering the amount of sensitive information printers can handle – from financial documents to personal records – the risk of malicious actors intercepting this data should not be underestimated [Insight IT, 2024].

> *"One in 16 home wi-fi routers tested vulnerable to default password attacks."*
>
> **Paul Bischoff**, Tech Writer, Privacy Advocate, and VPN Expert

Default Passwords

Factory configurations of many systems, devices, and appliances often include default passwords, that are typically the same for all products from the same vendor or the same product line. Such passwords are intended for initial testing, installation, and configuration activities. However, as soon the system, device, or appliance is ready to be used, you should change the default password with a stronger (and unique) one, and change the default username too.

The SANS Institute included default passwords in its "Ten Most Critical Internet Security Threats", not only because they are typically **weak**, **guessable** (some of the most common ones are *admin*, *guest*, *root*, and *1234*), and often **hardcoded**, i.e. directly written into the software source code. The main issue is that those weak, guessable, and often hardcoded passwords can be **easily discovered** by just about anyone. Vendor documentation, computer books, publications, and training courses for Information Systems (IS) auditors, as well as newsgroup discussions and hacker sites on the Internet have default password information that cybercriminals can use to infiltrate your network, and anything connected to it [GIAC, 2005].

Default credentials pose a serious security risk also because they usually carry high-level, **administrative privileges**. This means that whoever logs in with such credentials has the rights to install or remove software, change configuration settings, access any file, modify or erase data, etc.

I think you get the point.

If malicious actors break into one of your systems, devices, or appliances through default credentials, they can do pretty much anything they want. Moreover, once they are in, they can laterally move and hack additional systems, devices, and appliances, until your entire network is under their control.

The United Kingdom recently became the first country in the world to ban weak or easily guessable default credentials from IoT devices [Martin, 2024]. It is more than likely (and desirable) that other countries will follow suit in the near future. Irrespective of what governments and vendors may or may not do next, do not wait for them and change default credentials to all your devices. Now.

✓ **KEY TAKEAWAYS**

1. *Change default passwords and usernames*
2. *Default passwords are weak, guessable, often hardcoded*
3. *Default passwords can be easily discovered*
4. *Default passwords may carry high-level, admin privileges*

Tip #4:

Think before you click

Phishing

Despite the increasing popularity of chat apps like *WhatsApp* and social media like *Facebook*, *Instagram*, or *TikTok*, email remains central to digital communication, and its use continues to grow. By 2025, the number of email users worldwide is expected to reach a total of 4.6 billion, with the average number of sent and received emails per day believed to exceed 370 billion [Ceci, 2024]. Most of us exchange tens – if not hundreds – of emails every day. Some of them are legitimate emails from coworkers, business partners, relatives, friends. Others are notifications (e.g., a confirmation of an online order), reminders (e.g., to complete check-in for an upcoming flight), marketing offers, or (often unsolicited) advertisement. But a few of them – the phishing emails – are a real danger.

Phishing is a type of cyberattack where malicious actors send fraudulent emails that appear to come from a legitimate and reputable source. For example, the phishing email might pretend to be from your bank, and it could pressure you to immediately change your online banking log-in credentials, otherwise threatening to close your account. Sometimes the email may come – or *seem* to come – from a person you know (a relative, friend, coworker, etc.), either because their email account has been compromised, or the hacker uses an address similar to that of your acquaintance (e.g., *JohnDoe@yahoo.es* instead of *JohnDoe@yahoo.com*).

> *"If it looks phishy, it probably is."*
> **April Goode**, Management Consultant

The main goal of phishing is stealing sensitive information, typically in the form of usernames and passwords, credit card details, and bank account information.

In most cases, cybercriminals launch bulk phishing campaigns by sending out the same email to masses of people. In some instances, however, the attack can be specifically targeting you, using highly personalized information that hackers collected on the Internet or social media. This can happen, for example, if you share online too many details about destinations, dates, flights, and/or hotels of an upcoming trip you are going to make. The scammer may email you and pretend to be working for the airline you will be flying with, or for the hotel you will be staying at. The email will include details about date and destination of your flight, or check-in and check-out dates at the hotel, and you will be therefore inclined to think the email is legitimate. This highly personalized cyberattack is called **Spear Phishing**.

How can you evade a phishing attack?

The general advice is to always be skeptical of unsolicited and unexpected requests for sensitive information, think twice before clicking any links or attachments, and never reply to the email.

Here below some warning signs to look for when receiving a suspicious email:

- **Unexpected Email**. Were you expecting the email, or was it a surprise? If you know the sender, verify with them (possibly in person or on the phone, definitely not by replying to the email) whether the email is legitimate.

- **Unusual Timing**. Was the email sent on a weekend or a holiday, or in the middle of the night?

- **Unknown Sender**. It is not unusual to receive

emails from people we do not know, but doing so should put you on alert.

- **Suspicious Sender's Email Address**. Is there something in the sender's address that looks questionable? E.g., a misspelled domain such us *amazan.com* instead of *amazon.com*.

- **Vague Subject Line**. Is the subject line very specific, or something vague and generic as "*Your payment was declined*", "*Password Expiration Notice*", "*Delayed Shipping*", or "*Please review and sign your document*"?

- **Email Signature**. If there is a signature at the bottom of the email, does the information, formatting, and/or brand logo match the company guidelines of the sender? If possible, try to compare the signature to that of another sender from the same company.

- **Unusual Request**. Is the email asking you to provide personal or financial information such as log-in credentials, credit card details, or bank account information? If the email comes from a person you know, is the request (and the tone) consistent with the relationship you have with them?

- **Generic or Missing Salutation**. Is the email directly addressing you (by first and/or last name), or is the greeting missing or a generic *Hi*, *Hello*, *Dear customer*, etc.?

- **Tight Deadline**. Does the email try to create a false sense of urgency by calling for quick or immediate action? Any time you receive an email enticing you to respond quickly, take your time, think, and look carefully at the message.

- **Threatening Language**. Does the email claim dire consequences or drastic actions (e.g., closure of your account) for not responding immediately?

- **Fake Prizes.** Does the email say you won a sweepstake, contest, or lottery, and ask you to provide your financial information and pay for something like taxes, shipping and handling charges, or processing fees?

- **Suspicious Links or Attachments**. When in doubt, never click on a link nor download an attachment. Clicking can trigger a malware download or take you to a malicious website. Always hover your mouse over the link (without clicking it!), and check whether the URL looks suspicious (e.g., an alphanumeric sequence like *rth58az3.info*) or does not match the email's source (e.g., the email claims to be from *DHL*, but the destination URL has nothing to do with the *DHL* website).

- **Poor Spelling and Grammar**. Check whether the email contains typos, incorrect grammar, punctuation errors, or inconsistent formatting. In the past, this used to be one of the easiest ways to spot a phishing email. This is no longer the case today, mainly due to the evolution of Natural Language Processing (NLP) and the availability of Generative Artificial Intelligence (AI) chatbots like *ChatGPT*. With such tools, cybercriminals can generate error-free, very convincing, highly sophisticated phishing emails. According to CNBC, since the launch of *ChatGPT* in the fourth quarter of 2022, there has been a 1265% increase in malicious phishing emails [Violino, 2023].

Smishing

Smishing, or *SMS (Short Message System) Phishing*, is a form of phishing attack carried out over mobile text messaging instead of emails.

The main goal of smishing is the same as the more common email phishing: tricking unsuspecting victims into divulging sensitive personal information.

Typically, the scammer poses as a delivery company (e.g., *Amazon*, *DHL*, *FedEx*, *UPS*), a financial institution, a tech firm (e.g., *Google*), a government agency, or even one of your acquaintances. Messages may notify you of a missed packet delivery, a suspicious activity on your bank account, the compromise of your electronic device, an unpaid fine or tax, or they may pretend to be from a close friend or relative asking for money to cope with a 'financial emergency'. Other smishing attempts may claim that you won a prize or may ask you to submit a donation to a charity organization. The list is endless, since cybercriminals constantly devise new ways to defraud unsuspecting users.

Similar attacks can also be launched on *WhatsApp*, *Facebook Messenger*, *iMessage*, or *Microsoft Teams*, just to name some of the most popular chatting platforms.

The warning signs to look for when receiving a suspicious text message are similar to the ones you go through in case of a dubious email: unexpected message, unusual timing, unknown phone number, unusual request, generic or missing salutation, tight deadline, threatening language, fake prizes, suspicious links, poor spelling and grammar.

If you receive a suspicious text message, the best choice is to not respond (delete the message instead) nor click any links.

Clickbait

Have you ever run into social media posts – especially on *Facebook* – displaying eye-catching imagery, sensational or scary headlines, too-good-to-be-true deals, or tempting discounts? Chances are you ran into **Clickbait**.

Clickbait is online content that aims at tricking users into clicking on a link, image, or video, and drive traffic to a website or blog to increase advertising revenue [Paulyn, 2024].

Images are typically visually striking and evoking curiosity, fear, or outrage. Headlines often use hyperbolic statements like "shocking", "amazing", "heartbreaking", or "exclusive" to lure people into clicking. "You'll Never Guess What Happens Next" is a prime example of clickbait headline.

In most cases, clickbait is just silly, time-wasting, and annoying, but harmless – even though the risk of spreading misinformation and fake news should not be underestimated. However, in some instances, clickbait can direct unsuspecting users to malicious websites, with potentially nefarious consequences such as data breaches, malware installation, identity theft, and financial losses. In their posts, scammers may pretend to be well-known individuals, respected businesses, or even close friends or relatives of yours. It sounds like phishing, doesn't it? It is.

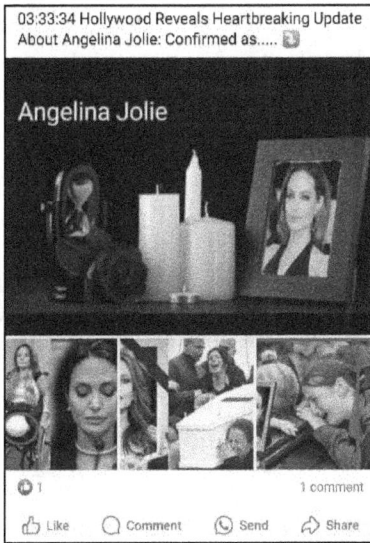

03:33:34 Hollywood Reveals Heartbreaking Update About Angelina Jolie: Confirmed as.....

Angelina Jolie

1 comment

Like   Comment   Send   Share

Clickbait on *Facebook* (example).

What makes clickbait a powerful cybersecurity threat is its ability to exploit the human appetite for the sensational, surprising, or unusual. It may stir up our curiosity, offer us instant gratification, or rely on our *Fear Of Missing Out*, or FOMO.

To a certain extent, dramatic or alluring titles are somehow expected and even tolerated in journalism and advertising, as long as they are not blatantly false or misleading. This is the case of **Fake News**, that are deliberately created to misinform and deceive readers, and used to influence political elections, discredit rivals, boost propaganda, stir up uncertainty, or spread panic. Fake news is one of the major concerns in today's globalized world, a world where a post or a video can easily go viral and be shared by millions of people.

Before closing this chapter, here are a couple of recommendations to cope with clickbait scams:

- **Be skeptical** of too-good-to-be-true stories, enticing offers, and provocative narratives. If it seems over-the-top and unbelievable, it is likely a scam.

- **Never click on a link if you do not know where it leads to**. Hover over the link with your mouse to see the destination URL.

- **Keep your anti-virus software up to date**. This is a good practice to combat many cyber threats, not only clickbait scams.

✓ **KEY TAKEAWAYS**

1. Be vigilant when you receive emails or messages, and when you are online
2. Don't click links, don't download attachments
3. Never reply to suspicious emails or text messages
4. Look for warning signs
5. Keep your anti-virus software up to date

Tip #5:

Be wary of unknown callers

Vishing

Similar to the phishing and smishing attacks we discussed in the previous chapter, **Vishing**, or *Voice Phishing*, is a type of scam in which cybercriminals contact their potential victims over the phone and try to convince them to share personal information or grant access to electronic devices.

Typically, scammers claim to be government representatives, pose as tech support, or impersonate bank employees. In the United States, they often pretend to be Internal Revenue Service (IRS) officers calling to collect taxes, threatening jail time if the victim does not pay their debt right away. In *Tech Support Scams*, a caller from a bogus helpdesk tricks the unsuspecting victim into paying a service fee or granting remote access to fix a device or software problem that does not exist. If vishers are given access to the victim's device, they may install malware or other undesired programs that can steal information or damage the data or the device itself.

Vishing scammers may also ask for charity donations, ensure big profits from an investment, offer free trials if you sign up for a new product, or promise low-cost – or even free – vacations. Those are just a few examples of the many vishing scams you may fall victim to.

"Be careful who you trust, the devil was once an angel."
Unknown

Some vishing attempts look like calls from unknown numbers, but it is not uncommon for hackers to create fake caller ID profiles or spoof existing ones, so that the phone number they are calling on appears to originate from a local area or trusted business, or even from one of your friends, coworkers, or family members.

Here below a couple of recommendations to avoid being vished:

- **Do not answer phone calls from unknown numbers**. Instead, let them go to voicemail (if you have one).

- **Do not post your phone number on social media**. In most cases, scammers collect the numbers of their potential victims from a compromised website or database, but sometimes numbers are spontaneously shared online by incautious users.

- **Do not reply to text messages from unknown numbers**. Scammers typically send SMS to a mass of random numbers, with a request to reply to the message. If the receiver does so, the sender gets confirmation that the number is in use and is therefore a potential target.

- **Pay attention to the caller's tone**. Scammers may use an unpolite or impatient tone to create urgency or threaten you. That is not how employees from the organizations they claim to represent would behave.

- **Listen for unnatural pauses or robotic-sounding speech**. Even though some vishing attempts are carried out by human beings, most of them are automated phone calls, or *robocalls*, delivering pre-recorded messages.

- **Avoid sharing personal or financial information over the phone**. If the caller is asking for your log-in credentials or your financial information, it is most likely a scam.

- **Hang up immediately if the caller sounds suspicious**. You can later verify the call's legitimacy by contacting the company or the individual through their official channels.

Audio Deepfakes

One additional advice is simply **not to speak**. If you ever pick up a call from an unknown number, stay silent. Do not engage. Do not ask "*Who is this?*" or say "*Stop calling me*". If you speak, you may fall victim to a *Voice Cloning Scam*, also known as *Audio Deepfake*.

In this phone scam, fraudsters steal a recording of the voice of their victims and use it to fake their identity. This is done using Artificial Intelligence (AI) voice generators, sophisticated software that is capable of emulating human speech with utmost precision, down to accent, pace, tone, pitch, style, and inflections of the individual being cloned.

Several AI voice cloning tools are available on the market, some of them even for free. *Descript*, *Fliki*, *Murf AI*, *Play HT*, and *Resemble AI* are reported to be among the best ones [Elegant Themes, 2024].

The more speech data these tools are exposed to, the better they can imitate a person's voice. Even a voice sample of a few seconds is enough to train an AI tool to mimic the speech pattern of an individual. Scammers can then use the cloned voice to stage a fake emergency to extort money from friends or family members of the cloned person.

One useful practice to combat voice cloning scams is to create a **safe word** (or phrase) with your loved ones that only you and they would know. In case any of you receive a suspicious call or voice message, you can use this safe word to verify each other's identity [Guarino, 2024].

Wangiri

Another type of phone scam is the so-called **Wangiri** (also known as *Ping Call* or *Missed Call* fraud).

Wangiri is a Japanese word that means "*one (ring) and cut*" [Guha, 2024]. This scam relies upon the innate curiosity of human beings. Swindlers make their potential victim's phone ring for just a few seconds, and then immediately hang up. If the recipient returns the call, they are charged high premium or international rates, with the scammers pocketing a share of the related fees. This scam is known to target both mobile and landline phones.

Fraudsters use various techniques to keep their victims on the premium rate call for as long as possible. They may use recorded messages, put the caller on hold, or even have people manning the line to keep their target talking or listening. The longer the call, the more money the scammers ultimately make.

The main advice in case of Wangiri scams is to **never call back an unknown number**, especially if it is an international one. Instead, look up the number online (e.g., on www.truecaller.com) to see if it has been reported as a scam.

Whatever phone scam you may be the target (or victim) of, we recommend **reporting the caller** to your phone service provider (e.g., *Vodafone* or *AT&T*). This will prevent other subscribers from being attacked by the same swindler.

✓ KEY TAKEAWAYS

1. Don't answer phone calls from unknown numbers

2. If you answer the call, stay silent as much as possible

3. Hang up immediately if anything sounds suspicious

4. Don't call back an unknown number

5. Don't reply to text messages from unknown numbers

6. Don't share your phone number online

7. Don't share personal or financial information in a call

8. Report the caller to your phone service provider

9. Agree on a safe word with your loved ones to verify identity

Tip #6:

Use caution with QR Codes

A **QR (Quick Response) code** is a 2D matrix barcode invented by the Japanese company Denso Wave in 1994, and now widely used all around the world for advertising, marketing, entertainment and transport ticketing, restaurant ordering, product labeling and tracking, electronic payments, interactive learning, or simply information sharing. The list is almost endless.

The COVID-19 pandemic outbreak accelerated even further the rise of QR codes as businesses switched to contactless payments and menus [Jogi, 2023a and 2023b]. According to Forrester Research, over half of online adults in France, the UK, and the US reported that they used QR codes [Carielli, 2023].

QR code design – usually a pattern of black squares on a white background – was modeled on the pieces from *Go*, a board game very popular in Asia. *Position markers* – the 3 black squares, one each in the upper left, upper right, and lower left corners of the QR code – enable scanning devices to immediately detect their presence.

Example of QR code.

In the majority of everyday cases, QR codes are simply a convenient replacement for a Web address, saving users the time and effort of typing a long URL.

But convenience comes at a price, and the price is **Quishing**, or *QR Code Phishing*.

Cybercriminals can create QR codes that take to malicious websites or apps – which often replicate well-known, legitimate ones –, where unsuspecting victims are tricked into downloading malware or entering sensitive information like log-in credentials or credit card details.

A recent analysis revealed that nearly 2% of all scanned QR codes were malicious [keepnet, 2024].

- In January 2022, scammers placed fraudulent QR code stickers on parking meters in Texas, redirecting victims to fake payment pages.

- In May 2023, a woman scanned what seemed to be a legitimate QR code on a tea shop menu. This triggered the download of a malicious application that granted cybercriminals access to sensitive information, including the woman's banking credentials.
- In December 2023, a fake QR code was used in a *Facebook Marketplace* transaction, tricking the victim into entering their log-in banking credentials on a malicious website.

The ones above are only a couple of examples of how quishing attacks can lead to financial theft, identity fraud, and malware distribution.

> *"I don't scan QR codes, and neither should you, especially if you care about cybersecurity."*
> **Morey Haber**, Forbes Council Member

Here below a few points to consider when dealing with QR codes:

- **Do not scan QR codes if they come from suspicious or unknown sources**. This is the case of codes you may find on random stickers, flyers or posters in public places, as well as those in unsolicited emails, texts, social media posts, or unfamiliar websites. Visual clues like customized design with company logo or branding usually confirm that the code is authentic.

- **Make sure QR codes have not been tampered with**. As demonstrated in the real-life quishing attacks above, threat actors can replace legitimate QR codes in menu holders (it is safer to ask for paper menus, if available), or place a sticker with a malicious QR code on top of the original code [Stickley, 2023].

- **Verify the destination URL** to confirm it is the intended site and looks authentic. Malicious domains are often similar to the legitimate URL but with typos or misplaced letters.

- **Secure websites include HTTPS** (*HyperText Transfer Protocol Secure*) in their Web address, not HTTP [Puri, 2024]. They also appear with a padlock symbol in the URL bar. *Google Chrome* and other browsers flag all non-HTTPS websites as 'not secure'. Visiting only HTTPS websites is a good practice for *all* online activities, not only when scanning QR codes.

- **Never download an app from a QR code**, use your phone's app store instead.

- **Use only trustful QR code scanning apps**. In 2020, hackers created fake QR code scanning apps that, once installed, downloaded malware. Most smartphones today have built-in QR code readers with security checks in their camera apps. Use those.

✓ KEY TAKEAWAYS

1. *QR codes are everywhere, so are the risks*
2. *Don't scan QR codes if you don't know their source*
3. *Make sure QR codes haven't been tampered with*
4. *Verify the destination URL*
5. *Secure websites include HTTPS and a padlock*
6. *Never download an app from a QR code*
7. *Use only trustful QR code scanning apps*

Tip #7:

Don't trust virtual friends

Privacy

Facebook, Instagram, Twitter, TikTok, LinkedIn, Xing, WhatsApp, Meetup, Snapchat, YouTube... the list is almost endless, as it could be the number of your virtual friends, followers, viewers, contacts, and connections. In today's digital world it is easy to stay in touch with people living literally on the other side of the planet.

According to the *Six degrees of separation* theory, based on Milgram's small world study [Milgram, 1967], any 2 people in the world are six or fewer social connections away from each other. Regardless of whether the links in the chain are exactly six or another number, are the people we virtually connect to the ones they *claim* to be? Have we met all of them in real life? Probably not.

It is not uncommon to receive contact requests on social media from people you genuinely believe you have never heard of before. Though many of them are probably legitimate requests from people you actually met but forgot about, or who have a valid (business) reason to contact you, chances are that a fair portion of them are scammers trying to hook you and snoop into your online activity (and – eventually – your private information).

> *"Amateurs hack systems, and professionals hack people."*
> **Bruce Schneier**, Cryptographer and Security Expert

While it is advisable not to accept a request unless you receive evidence that you really know each other (ask them to remind you when, where, and under which circumstances the two of you met), there are social media settings that are designed to protect your account. Here below some instructions on how to **adjust privacy settings** in *Facebook* and *WhatsApp*, two of the most popular social apps today.

Facebook
1. Click the **Account** icon on the top right of the *Facebook* homepage
2. Click ✿ (**Settings & privacy**), then click **Settings**
3. Under **Audience and visibility**, click the option you

want to change. Among other things, you can select how people may find and contact you, who can follow you, and who can tag you

WhatsApp
1. Tap ⋮ , then **Settings**, then **Privacy**
2. Tap the privacy setting you would like to change, and select *Everyone*, *My Contacts*, *My Contacts Except...*, or *Nobody*. Among other things, you can silence unknown callers and select who can see your profile photo, about information, status updates, etc.

In some cases, scammers hack social media accounts or set up fake profiles using personal information and pictures stolen from unsuspecting users. This is what I witnessed a couple of years ago, when I received an unexpected request for money from a coworker (and *Facebook* friend). The request sounded so awkward (Why contacting me on *Facebook* instead of meeting in person or giving me a call? We were colleagues, not close friends: why asking me for financial help?) that it did not take me long to spot the fraudster. This story had a happy ending, but too many others end up with victims sending money or sharing sensitive information with cybercriminals.

Dating Scams

A distinctive type of scam often takes place on dating apps like *Tinder*, *OkCupid*, or *Bumble*. In today's frantic world more and more people meet their partners on the Internet – even more since the COVID-19 pandemic. While some find their long-awaited sweetheart (and I personally know of a few successful couples who started their love story in the virtual world), many are victims of **dating scams** (also referred to as **romance scams** or **sweetheart scams**).

In this type of fraud, cybercriminals create a fake online profile (often using pictures stolen from attractive individuals) and start a relationship with those looking for true love. This process of luring someone into a relationship by means of fictional online personas is known as *Catfishing*. Once scammers win the trust of their victims and have their hooks in them, they usually ask for money or financial information. In most cases, malicious actors justify their requests by pretending they – or one of their family members – are sick, injured, or in prison, or by enticing their prey to put money into what they swear to be a safe and profitable investment. According to the U.S. Federal Trade Commission, victims lost $1.14 billion to dating scams in 2023, with median losses per person amounting to $2,000 [Solá, 2024]. And that says nothing about the emotional damage inflicted by romance scammers on their preys. Almost three-quarters (73%) of the victims were men, with vulnerable people like older widows or divorcees being the main targets.

Here below some of the warning signs to watch for when romancing online [Masjedi, 2023]:

- **Refusal to meet in person or appear on video chat**. Scammers make up excuses about why they cannot meet you (e.g., they claim to live in a foreign country or pretend their camera is broken or their Internet connection is slow): They may even ask you money for travel expenses to come out and meet you.

- **No digital footprint**. Since online scammers set up fake profiles using stolen information and photos from real people, they do not appear on social media platforms, school listings, and so on. Do a simple online search or use a tool like *Google Lens* to collect information about them and check whether their photos are authentic.

- **Isolation attempts**. Fraudsters discourage you from talking to family and friends about your new romantic interest, and they insist on your silence.

- **Request to leave the dating app** and move to messaging apps like *Snapchat*, *Telegram*, or *WhatsApp*. That is a way for imposters to access your personal information (like phone number or email).

- **Too good to be true – especially in photos**. Scammers often use stolen profile pictures that are too attractive or magazine-quality, and do not have many of them available. Also, they pretend to be interested in the same things you are, trying to move the online relationship forward very fast.

- **Anonymous payment methods**. Con artists usually demand hard-to-track payments, like wire transfers, cryptocurrencies, or pre-loaded gift cards.

Dating scams are some of the more insidious (and vicious) cybercrimes because they prey on the victim's emotions and vulnerability. If you realize you are the target of such a scam, here are a few actions you may want to take [TWU, 2024]:

- Immediately stop communicating with the

imposter.

- Take screenshots of email address, phone number, or any other contact information of the scammer you may have.

- Report the incident to local law enforcement.

- Alert the website, social media platform, or app where the two of you met. They may have information about the scammer that can help with the investigation, or at least they can delete the fraudster's profile, thus protecting other potential victims.

- Regularly check bank and credit card statements for any unusual activity.

Financial Sextortion

A recent variant of dating scams is the so-called **Financial Sextortion**. This new scam – mainly targeting teenage boys and young men – exploded over the past three years to become the fastest growing cybercrime.

The way sextortion works is relatively simple and straightforward. Scammers create fake profiles with very attractive pictures (usually girls), then send out flirtatious messages to their targets on social media, which culminate with a request to the victim to share sexually explicit photos or videos, possibly showing both face and genitals. Once the victim hits the send button, the threat begins. The offender demands a ransom – to be paid typically through cryptocurrency or gift cards – to keep the images private, threatening to publish the compromising material or share it with the victim's family members and social media contacts. According to the FBI, between January 2021 and July 2023, at least 20 teenagers committed suicide when threated with compromising material [Maslin Nir, 2024].

In some cases, cybercriminals blackmail their victims by claiming they have hacked their smartphone or computer and are now in possession of compromising footage (which they are not). To be on the safe side, many victims choose to pay the ransom.

A couple of recommendations to avoid falling victim to sextortion [NCIS, 2016]:

- Cover webcams and turn off electronic devices when you are not using them.

- Do not engage in sexually explicit activities online, such as posting or exchanging compromising photos or videos.

Revenge Porn

Another form of sexually oriented cybercrime is **Revenge Porn**, or **Non-Consensual Pornography** (NCP). This is the distribution of private, sexually explicit pictures or videos of individuals online without their consent, and with the objective to humiliate or embarrass them. In the United States, as many as 1 in 12 adults have been the targets of revenge porn, with the vast majority of victims being women [Glassner, 2021]. This form of cyber abuse has led to job losses, psychological traumas, and – in some tragic cases – even suicide.

Here below are some actions you may take if you are a victim of revenge porn [Cybercrime Support Network, 2022]:

- Contact the website or social media platform where images and/or videos are posted and try to get them removed.

- Report the perpetrator to the police. Revenge porn is a crime in many countries, including Germany, Italy, the United Kingdom, and most of the United States.

One final word of caution before closing this chapter: crime does not stop online. Serial killers and sexual predators exist and are out there. If you are planning to meet your virtual date or new social media friend in the real world, schedule your appointment in a public place and let someone else (friends or family members) know where you are going and whom you are meeting with. And have your smartphone with you in case an emergency call is needed.

✓ KEY TAKEAWAYS

1. Don't accept contact requests from people you don't know

2. Adjust privacy settings in the social apps you use

3. Be alerted if you receive unexpected requests (e.g., for money) from your social media contacts

4. Watch for warning signs when romancing online

5. If you are the target of a dating scam, take action

6. Cover webcams and turn off devices when not in use

7. Don't engage in sexually explicit activities online

8. If you are victim of revenge porn, report the crime to the police and contact the website or social media platform

9. Meet your virtual contacts in a public place and ensure friends or family know where you are and with whom

Tip #8:

Protect your Digital Footprint

Any time we are online to book a hotel, post on social media, download an app, activate an account, purchase an item, subscribe to a newsletter, fill out a survey, check the weather, or order food– just to name a few of our everyday activities in today's digital world – we leave a trail of digital breadcrumbs behind us, similar to footprints on wet sand. This is our **Digital Footprint**, and it portrays our interests, hobbies, beliefs, and habits. However, while waves will eventually wipe out real footprints, digital ones might be permanent.

Any information we share online can be tracked and analyzed by marketing companies, credit card providers, advertisers, law enforcement agencies, and other organizations to learn about who we are and what we do, and create personalized offers and profiles accordingly. Our online presence defines who we are, or at least who we *appear* to be.

We should always remember that while we look up all type of information online, the Internet is looking right back at us. Anything we do in the digital world contributes to our reputation, to our "personal brand", and – what is even worse – may expose us and those around us to cybercrime, like targeted phishing attacks, account takeover frauds[2], extortion scams, and identity theft.

Here below are some tips to limit the potentially negative impact of our digital footprint [Morgan Stanley, 2023].

- **Do an online search of your name**.
  This will allow you to see what information about yourself is readily available on the Web. If you find sensitive data being disclosed (e.g., your home address or phone number), or you come across information that is incorrect, misleading, or inappropriate, you should contact the site administrator and request the material to be removed.

- **Set an alert.**
  With an alert about you, you will receive a notification any time your name pops up online. You should try different search engines, such as

---

[2] *Account Takeover Fraud* (ATO) is when scammers gain access to your online accounts (social media, email, online banking, etc.).

Google or *Bing*. As an example, you can set an alert in *Google* as follows:

a. Go to google.com/alerts
b. Select the option *Me on the web*
c. Click *Create alert*
d. If desired, you can change settings and select frequency of notifications, language, part of the world you want info from, etc.

- **Restrict your privacy settings.**
  This can help reduce your digital footprint. As an example, you can adjust your privacy settings in *Facebook* as follows:

a. Click the **Account** icon on the top right of the *Facebook* homepage
b. Click ⚙ (**Settings & privacy**), then click **Settings**
c. Under **Audience and visibility**, click the option you want to change for privacy. Among other things, you can select how people may find and contact you, who can see your posts, stories, and reels, who can follow you, and who can tag you.

Moreover, any *Facebook* post allows you to select the audience you want to share it with (e.g., *Public, Friends, Friends except..., Only me*).

- **Be cautious with what you post.**
  Be careful with any photos, videos, comments, and likes you share online. Not only what you post may be tied to you forever, but it might give scammers personal details to steal your identity, or that could be used against you, as we have seen with *Spear Phishing* in our Tip #4. Also, never post anything you might regret in the future, such as aggressive, bullying, racist, or purely dumb comments. With a touch of a button, you can ruin your chances at attending a university, lose your job, or ruin a relationship.

> *"In the future, your 'digital footprint' will carry far more weight than anything you might include on a resume."*
> **Chris Betcher**, Educator & Consultant

- **Limit app permissions.**
  When you install an app on your smartphone, you will likely see a request to grant the app access to your camera, microphone, GPS location, contacts, calendar, or other sensitive information that could

be used maliciously. Apps may legitimately require some permissions to function properly (for example, *Google Maps* needs access to your location to give you precise directions), but you should avoid any permissions that are not strictly necessary. A good practice is to turn off location tracking when not using apps that require it to function.

- **Delete unused apps and accounts.**
  You can reduce the risk of having your personal information being misused or stolen by deleting any apps you no longer use, as well as all your old and unused accounts, including email accounts. Simply deleting an app on your phone does not delete the associated account. Be selective about opening new accounts, too. If you are given the option, check out as a guest with a retailer instead of creating an account. Similarly, unsubscribe to services and newsletters you no longer need.

- **Avoid linking accounts.**
  Some apps let you sign up by using the log-in information you have with another company, such as *Facebook* or *Google*. If you do so, these apps will be granted access to even more information about you and your online activity.

a. To see apps connected to your *Facebook* account, log into *Facebook* using a computer, click your profile picture on the top right, then click **Settings & privacy→Settings→Apps and Websites**.

b. To see apps connected to your *Google* account, sign in, then go to https://myaccount.google.com/permissions.

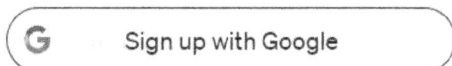

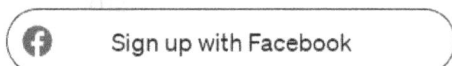

G   Sign up with Google

f   Sign up with Facebook

Signing up with *Google* or *Facebook*.

- **Use Private Browsing.**

  Also known as *incognito mode*, *stealth mode*, or *private mode*, it is a feature available in most browsers (e.g., *Google Chrome*, *Firefox*, *Safari*) that allows users to browse the Internet without savings cookies, temp files, and a history of the pages they visit. Private browsing is available on computers as well as on smartphones. As an example, in *Google Chrome* it can be activated by tapping ⋮ , and then selecting **New Incognito window**.

✓ **KEY TAKEAWAYS**

1. Search your name online
2. Set an alert
3. Restrict your privacy settings
4. Be cautious with what you post
5. Limit app permissions
6. Delete unused apps and accounts
7. Avoid linking accounts
8. Use Private Browsing

Tip #9:

Think twice before you plug it in

Removable Media

Removable Media is any type of portable storage device that can be inserted into a computer, printer, or other piece of hardware, and later removed while the system is running. In addition to data, such devices can store audio, video, image, and game files.

There is a wide range of removable media on the market today, with different storage capacities, data transfer speeds, and physical connectors: from old-fashioned magnetic tapes and floppy disks to optical disks (CDs, DVDs, BDs) and external hard drives, to USB sticks and SD cards.

Removable media offers numerous benefits:

- **High Portability**, thanks to large storage capacities, small sizes, and light weights.

- **Easy and Fast Data Transfer**, with no need for complex setup nor Internet connectivity.

- **Affordable Cost**, which is constantly declining over time.

However, these benefits can be outweighed by the substantial security risks of removable media, and the potentially devastating consequences to which such risks could lead, namely:

- **Malware**: since portable devices are frequently used to transfer data between different systems bypassing network security measures, they can easily become carriers of malicious software.

- **Physical Loss or Theft**: most removable media are small and easy to lose or get stolen.

- **Physical Damage**: dropping the device or exposing it to extreme temperatures can compromise the data stored on it.

- **Data Exfiltration**: removable media can be used by cybercriminals to steal sensitive data from another digital device (e.g., a computer).

"Security is always excessive until it's not enough."

Robbie Sinclair, Cybersecurity Consultant & Recruiter

One of the trickiest ways to use removable media to cause harm is the **USB Drop Attack**. In its most common form, this hacking method involves intentionally leaving a malware-laden USB drive in a public space (e.g., a parking lot), with the hope that someone will be lured to pick it up and plug it into their computer.

Here below some best practices to mitigate removable media security risks [Verizon, 2024]:

- **Do not plug unknown devices into your computer**. Beware of any removable media you may accidentally bump into (you still remember *USB Drop Attacks*, don't you?) or that is handed over to you (e.g., as a gift at a conference).

- **Buy removable media only from trusted sources**. Be careful where you shop, especially if you do it online. It is not uncommon to receive removable media (especially USB sticks) infected by malware.

- **Scan external devices for malware before accessing them**. Ensure that the host device has an up-to-date anti-virus/anti-malware program installed.

- **Disable AutoPlay and AutoRun features**. AutoRun viruses are programs that automatically execute when an infected external device is plugged into the host. The details of how to disable such features are beyond the scope of this book, but there is plenty of documentation available online (e.g., [Landesman, 2022]).

- **Set up a (strong) password** to prevent unauthorized access and protect any sensitive data stored on the portable device.

- **Encrypt sensitive data stored on removable media**. Encryption ensures that even if the device falls into the wrong hands, data remains unreadable without the decryption key. Pre-encrypted flash drives are available on the market (e.g., on *Amazon*). They are typically more expensive than regular flash drives, but the extra security can be worth the money.

- **Do not leave removable devices unattended** or in places where they could be easily stolen.

- **Clear all data on the device once you no longer need it.**

Mobile Devices

Mobile devices like smartphones, tablets, and smartwatches are a special type of removable media. Their widespread and ever-increasing use has led to the installation of free USB charging stations in public places all around the world – airports, hotels, shopping malls, event venues, etc. Chances are you may have used a couple of those yourself while traveling or shopping. Well, be aware that juicing up your electronic device at free USB port charging stations could have unfortunate consequences: you may become a victim of **Juice Jacking**.

Juice Jacking is a security exploit in which devices are compromised when connected to an infected USB port, or use an infected cable attached to a public charging station. While the mobile device is charging, data is transferred to the cybercriminal, or malware uploaded to the device. This type of exploit works because the same USB connector that charges the device also transfers data. The amount of data an attacker may be able to extract depends on how long the device is left plugged into the compromised port or cable.

Here are a couple of tips to avoid falling victim to Juice Jacking [Trevino, 2023]:

- **Avoid public charging stations**. If you must charge your phone, use a wall socket instead.

- **Carry external batteries or power banks** when on travel, as well as **your own charger**. Never use a lost or unverified charging cable.

- **Use a USB condom** (a.k.a. USB Data Blocker). This is a cheap, protective attachment that you plug into the charger side of your USB cable before connecting it to a public USB port. The condom allows power to pass through but not data, thus preventing files from moving between your device and the hacker's.

- If you plug your device into a public USB port and a prompt appears, always **select "*charge only*" and "*don't trust*"**. By doing so, charging will be allowed but data transfer will not. Unfortunately, this does not stop attackers from uploading malware to your device.

- **Power your device off** before plugging it in, even though the USB port may still connect to the flash storage in the device.

Although Juice Jacking has been demonstrated to be technically possible as a proof of concept at the DEF CON hacking conference in Las Vegas in 2011, so far there are no confirmed instances of it occurring [FCC, 2023]. However, precisely because it *is* possible, it is better to assume it may happen and take reasonable precautions.

✓ KEY TAKEAWAYS

1. *Don't plug unknown devices into your computer*

2. *Scan external devices for malware before accessing them, disable AutoPlay and AutoRun features*

3. *Buy removable media from trusted sources, protect it with passwords and encryption, don't leave it unattended, and clear all data once you no longer need it*

4. *Avoid public charging stations, use wall sockets instead*

5. *Carry external batteries, power banks, and your own charger when traveling*

6. *When using a public USB port, power your device off, use a USB condom, and select "charge only" and "don't trust" if prompted*

Tip #10:

Avoid using public Wi-Fi

Public Wi-Fi

In today's highly-digital world, many people are longing to stay connected 24/7 for both leisure and work activities, and **public Wi-Fi** – widely available in hotels, restaurants or cafes, airports, public transit, malls, retail stores, libraries, etc. – offers them access to the Internet for free.

But nothing comes for free.

"The biggest threat [with public Wi-Fi] is your data, traffic and identity could be completely exposed."

David Lee, Product Manager at Norton

In this case, the price you have to pay are the risks your data and your software will be exposed to, especially when connecting to public Wi-Fi networks than do not require a password. Since anyone can join a public Wi-Fi network, you do not know who is watching or intercepting your online data.

According to a recent study by Forbes [Haan, 2023], 43% of the people surveyed had their information compromised while using public Wi-Fi, with the highest percentage (25%) reporting it happened at a cafe or restaurant.

Man-in-the-Middle Attacks

Public Wi-Fi networks come with significant risks like data interception, malware, and identity theft [Buxton, 2024]. One of the most prominent threats is the risk of **Man-in-the-Middle**[3] (MitM) attacks. This is a type of cyber-attack in which cybercriminals secretly intercept messages between two parties (for example, between you and your bank), and capture and/or manipulate sensitive personal information like usernames and passwords, bank account details, or credit card numbers. The term "man-in-the-middle" refers to malicious actors essentially operating as "middlemen" between the sender and the recipient of information. Online banking and e-commerce sites

---

[3] Called also *Meddler-in-the-Middle*, *Monster-in-the-Middle*, *Machine-in-the-Middle*, *Monkey-in-the-Middle*, etc. for more inclusiveness.

are the prime targets of such attacks [Yasar, 2022].

Sometimes threat actors perform MitM attacks by setting up **Rogue Access Points**. These are fake wireless access points, often with a name very similar to a legitimate one (e.g., "*Hilton Honors 5G*" instead of "*Hilton Honors*", or "*Starbucks High-Speed WiFi*" instead of "*Starbucks WiFi*"). Unaware victims connect to the rogue access point instead of the legitimate public network, thus allowing the attacker to steal sensitive information or spread malware. Our recommendation, if you need to use public Wi-Fi, is to always confirm with local staff the exact name (and hopefully password) of the network you are planning to connect to.

Virtual Private Networks

The best way to protect your data when connecting to public Wi-Fi networks is to use a **Virtual Private Network (VPN)**. A VPN encrypts Internet connections and online data transfers – including your browsing history – thus keeping your data and identity hidden. Even if an attacker manages to access a network, they will not be able to access nor manipulate your data due to the encryption provided by the VPN.

There are many VPN services available on the

market, with subscriptions as low as $2-3 per month. In September 2024, *security.org* assigned their top 3 security scores to *NordVPN*, *Surfshark*, and *Private Internet Access VPN* [Vigderman, 2024].

In general, entering personal information such as passwords or credit card numbers is a hazard while connected to a public Wi-Fi network. However, if you do access any of your accounts, always remember to **log out** when you are finished.

✓ **KEY TAKEAWAYS**

1. *Public Wi-Fi networks pose major cybersecurity risks*

2. *Confirm with local staff name and password of public Wi-Fi network*

3. *Use a VPN when connecting to a public Wi-Fi network*

4. *Avoid entering personal information when using a public Wi-Fi network*

5. *Remember to log out when finished*

Tip #11:

Keep a close eye on Bluetooth

Bluetooth is a wireless communication technology that uses radio waves for exchanging data between different devices within a short distance, typically around 30 feet (10 meters).

The name dates back to the Danish King Harald "Blåtand" Gormsen and his rotten, bluish-colored tooth. The Bluetooth logo itself merges the Hagall (✳) and Bjarkan (ᛒ) Scandinavian runes, Harald's initials. Just as King Harald united Scandinavia in the tenth century, major industry players in the late 1990s wanted to develop a common standard that would allow all devices to communicate wirelessly over short distances.

Bluetooth technology has been an essential part of our lives for over two decades, mainly because it is free, easy to use, and independent of manufacturers and operating systems. Most likely you already used Bluetooth a couple of times in your life, maybe to pair your smartphone with a speaker, smart TV, fitness tracker, vehicle, or another smartphone. You may have used Bluetooth to turn your phone into a Wi-Fi hotspot and share its internet connection with your laptop (this is called *Hotspot Tethering*). You may have a computer that uses Bluetooth to connect to a wireless keyboard, mouse, or headset.

> *"New security loopholes are constantly popping up because of wireless networking. The cat-and-mouse game between hackers and system administrators is still in full swing."*
>
> **Kevin Mitnick**, Computer Security Consultant & Hacker

However, widespread use also means high risk of security breaches, typically targeting smartphones, laptops, and tablets. There are many ways cybercriminals can exploit Bluetooth technology. Discussing the details of the different types of Bluetooth-related cyberattacks and how they work is beyond the scope of this book. What is important for you to know is that malicious actors can use a Bluetooth connection to steal sensitive data on a device (*bluesnarfing*), send unsolicited messages with malicious links or attachments (*bluejacking*), gain full control of a device (*bluebagging*), cause a device to crash or become unresponsive (*bluesmacking*), or eavesdrop on conversations and phone calls that take place inside a vehicle (*car whispering*) [Stouffer, 2022]. Scammers also found ways to hack Bluetooth technology to remotely unlock and operate vehicles with keyless entry.

Here below a couple of security tips to protect your device and data from potential threats when using Bluetooth [Alvarez T. G., 2024]:

- **Turn it off when not in use**, and especially when you are in highly trafficked areas like airports, malls, and coffee shops. Disabling Bluetooth may also increase your device's battery life.

- **Avoid pairing devices in public places**, especially when you do it for the first time. Malicious actors could hijack the pairing process and connect to your device.

- **Never accept pair requests from unknown devices**. If an unknown device is trying to connect to yours, most likely it is someone attempting to gain unauthorized access.

- **Do not share sensitive information over Bluetooth**, especially if you are in a public area.

- **Delete unused Bluetooth connections**. This will ensure your device will not automatically connect to a paired device that got compromised. It is also a good practice to disconnect from a device after you have finished using it rather than leaving it connected.

✓ **KEY TAKEAWAYS**

1. Turn Bluetooth off when not in use

2. Avoid pairing devices in public places

3. Never accept pair requests from unknown devices

4. Don't share sensitive information over Bluetooth

5. Delete unused Bluetooth connections

Tip #12:

Update now, not later

Vulnerabilities

Cybercriminals never stop looking for ways to gain unauthorized access to their victims' digital devices. *Our devices.* One of the most common ways to deliver a successful attack is by exploiting weaknesses in information systems, also known as vulnerabilities.

Vulnerabilities are flaws in hardware or software that allow threat actors to cause harm, like spreading malicious software or exfiltrating sensitive data. Much like a burglar trying to break into a house, cybercriminals always look for the easiest way to get in, and an open vulnerability is equivalent to leaving a door unlocked.

Once a vulnerability is discovered, it is registered as a **CVE** (*Common Vulnerability and Exposure*) and assigned a **CVSS** (*Common Vulnerability Scoring System*) score to qualitatively measure its severity [NIST, 2024]. CVSS scores range from 0 to 10, with 10 indicating highest severity. The higher the CVSS score, the easier it is for hackers to exploit the vulnerability.

Patches

Software developers are constantly creating **patches** for their products to fix known vulnerabilities and improve functionality. Such patches become available through operating system and app updates. Despite the temptation to delay them, it is paramount you install updates – including anti-virus and anti-malware software – as soon as your device prompts you to do so. Ignoring or delaying updates leaves you vulnerable to attacks.

Software Update

iOS 13.3 is now available for your iPhone.

Close | Details

Software Update Prompt (iPhone).

Keep in mind that once a vulnerability is known to developers, it is also known to cybercriminals, who will try to exploit it before you update your devices. Since this is a never-ending process, you need to continually download the latest patches and keep your software up to date.

Automation & Rebooting

The easiest way to stay on top of this is by turning on or confirming **automatic updates** where possible [ACSC, 2024]. An automatic update is a feature that checks for, downloads, and installs itself in the background without requiring any action on your part.

> *"60% of data breaches are caused by a failure to patch. If you correct that, you've eliminated 60% of breaches. And I didn't even have to say AI or Blockchain! See how that works?"*
> **Ricardo Lafosse**, CISO at Kraft Heinz

Regularly **rebooting** your device – we recommend at least once a week in case of personal computers and smartphones – is crucial to fully integrate security updates into your system, thus optimizing its overall functionality and fortifying its defense mechanisms against the latest threats. This applies to desktops and laptops, smartphones, tablets, as well as many other digital devices, including Wi-Fi routers, smartwatches, and smart TVs [Hill, 2022]. The benefits of rebooting a device are not limited to applying security updates, but include installing new features, resolving bugs, freeing up memory, improving software performance, and reducing energy consumption [Northwestern University, 2024].

Other updates may require manual download and installation. Here below some easy-to-follow instructions to manually update apps on *Apple* and *Android* smartphones.

Apple
1. Open the **App Store**
2. Tap the **My Account** button or your photo at the top of the screen
3. Scroll down to see pending updates and release notes. Tap **Update** next to an app to update only that app, or tap **Update All**

Android
1. Open the **Google Play Store** app
2. At the top right, tap your profile icon
3. Tap **Manage apps & device**, then **Manage**

4. Tap the app(s) you want to update (the ones with an update available are labeled *"Update available"*)
5. Tap **Update**

A couple of additional recommendations before closing this chapter:

- **Consider replacing devices or apps that can no longer receive updates**.

- **Install updates only from official sources**, such as a trusted app store. Be extremely cautious in case you are prompted to install an update through a browser pop-up or an email, as these could be phishing attempts [CISA, 2024].

✓ **KEY TAKEAWAYS**

1. *Install updates as soon as they become available*
2. *Enable automatic updates where possible*
3. *Regularly reboot your devices*
4. *Consider replacing devices or apps that can no longer receive updates*
5. *Install updates only from official sources*

Tip #13:

Backup is your policy insurance

Ransomware

The *2023 Cybersecurity Ventures Cybercrime Report* predicts the cost of ransomware attacks to keep rising in the years to come, reaching around $265 billion per year worldwide by 2031 [esentire, 2024]. This means approximately half a million US dollars *per minute*!

Anyone with a device connected to the Internet is at risk, including government or law enforcement agencies, education and healthcare providers, critical infrastructure entities, large and small businesses, private citizens. *All of us.* 71% of organizations worldwide were reportedly affected by ransomware attacks in 2022 [Griffiths, 2024]. Years ago, a close friend of mine fell victim to a ransomware attack which encrypted – among other files – all pictures and videos on his smartphone.

But what is **ransomware**?

Ransomware is a form of malware designed to

block access to and/or steal digital data until a sum of money (*ransom*) is paid to the attacker. Payments are usually made in cryptocurrencies – mainly due to their anonymity and decentralized nature –, with Bitcoin accounting for 98% of ransomware extortion demands [Fuhrman, 2020].

Ransomware strategies usually fall in one of the following 3 categories [Channel, 2020]:

1. **Crypto Malware**: data on the victim's device gets encrypted and a ransom payment is demanded in exchange for the associated decryption key. Notable crypto malware strains are *CryptoLocker*, *Petya*, *WannaCry*, *Maze*, *Revil*.

2. **Lockers**: access to the victim's device – or part of it – is denied until a ransom is paid to restore it. Examples of lockers are *WinLock*, *Reveton*, *LockerPin* [Check Point, 2024].

3. **Doxware**: in this case, the malicious actor has stolen data from the victim's device and is threatening to publish it online if a ransom is not paid. The word "dox" comes from "documents", and refers to the act of stealing documents and other sensitive information from victims [Abed, 2023]. *Financial Sextortion*, discussed in Tip #7, may fall in this category.

Over 300 million ransomware attempts were detected in 2023 alone [Petrosyan, 2024]. Chances are you heard of some of the following attacks, as they grabbed headlines around the world:

- **WannaCry**: in May 2017, the *WannaCry* ransomware spread to over 200,000 computers in more than 150 countries. Notable victims included FedEx, Honda, Nissan, and the United Kingdom's National Health Service (NHS) [Cloudflare, 2024].

- **Snake**: in June 2020, the *Snake* malware forced automotive giant Honda to temporarily shut some of its production facilities around the world, as well as both customer support and financial services operations [Winder, 2020].

- **DarkSide**: in May 2021, *DarkSide* attacked Colonial Pipeline, the owner of a large system carrying fuel from Texas to the Southwest United States, causing gas supply shortages that impacted 17 states and Washington, D.C. for several days.

Ransomware attacks often start with **phishing**, which tricks victims into clicking on a link or downloading an attachment that contains the malware. Once the malware is on the victim's device, it encrypts data and/or steals sensitive information.

WannaCry Ransomware [Challita, 2017].

Paying the **ransom** is not recommended, for several reasons:

- First, by sending your money to cybercriminals you would only confirm that their ransomware works.

- Second, payment does not guarantee you will get the key you need to decrypt your data.

- Third, even if you receive the key and get your data back, there is nothing preventing the same malicious actors to hack you again and demand a new ransom.

Backups

The best way to cope with ransomware – at least with its crypto malware and locker types – is to have

backups available. Our digital devices are treasure chests full of private data: from family photos and videos to music collections, banking information, health records, and personal contacts. Storing all this information on a smartphone, laptop, or tablet comes with the risk of it being lost if no backup exists.

> *"Backups represent the last line of defense for everything from fat fingers to state-sponsored attacks."*
> **Rob Price**, Field Security Officer at Snow Software

Ransomware attacks are not the only reason why backups are needed. Many other unpredictable events can cause data losses, including [Morgan Stanley, 2024]:

- *Human error* (e.g., accidentally dropping your device, or spilling some liquid on it)
- *Hardware failure* (especially when exposing it to overheating or magnetic fields)
- *Accidental deletion*
- *Lost or stolen device*
- *Natural disasters* (fires, floods, earthquakes, etc.)

Many backup solutions are available on the market today. Some of them are free of charge, others require a subscription fee or a small investment to purchase backup hardware. Choose a solution that fits your needs and budget, and – important! – ensure it is reliable and secure.

Backup solutions can be subdivided into two main categories:

- *Cloud-based services*, like Apple iCloud, AWS Cloud Backup, DropBox, Microsoft OneDrive. As a reference, Google offers a paid subscription plan – *Google One* – starting at $1.99 per month.

- *Backup hardware*, mainly removable media like external hard drives, USB sticks, CDs, DVDs, or SD cards. As a reference, a SanDisk 128 GB USB stick can be purchased for about $10.

As Dr. Ullrich from SANS suggests, it is advisable to maintain not 2 but 3 copies of critical data: the one on the original device (smartphone, laptop, etc.), a cloud or online remote copy, and an offline remote copy (e.g., on an external hard drive) [King, 2023].

To conclude, a couple of suggestions for creating an effective backup strategy [Abed, 2023]:

- **Back up your data regularly**: the more frequently

you back up, the less data you will lose if something goes wrong.

- **Store backups offsite**, on the cloud and/or in a location different from your home or office. This will ensure that at least one copy of your data survives natural disasters, theft, ransomware attacks, etc.

- **Test your backups** to make sure they work correctly and can be easily restored if necessary.

- **Change your credentials** (usernames and passwords) after restoration, in case you fall victim to a ransomware attack.

✓ **KEY TAKEAWAYS**

1. *Any device connected to the Internet is at risk*
2. *Ransomware often starts with phishing*
3. *Paying the ransom is not the best solution*
4. *Backups are your insurance against data loss*
5. *Maintain 3 copies of critical data, at least 1 offsite*
6. *Back up your data regularly*
7. *Test your backups*
8. *Change credentials after data restoration*

CONCLUSIONS

Congratulations! You made it to the end!

I hope this reading helped you gain a deeper understanding of the risks any of us may face when using an electronic device. There is no doubt that technological progress produced countless benefits, but it also generated significant new threats in the digital – and sometimes even physical – world. Cybercrime is constantly growing, and the global cost of cyberattacks increases by 15% every year [Touchtidou, 2024]. When surfing the Web or tapping our devices, we need to be aware of the dangers, exercise caution, stay vigilant.

In this book, we discussed the importance of having strong and unique passwords for each one of our digital devices and accounts, implementing *Two-Factor Authentication* (*2FA*) wherever possible, and promptly changing default credentials, especially on *IoT* devices. We mentioned the risks of *Brute-Force* and *Credential Stuffing Attacks*, as well as the benefits of using *Password Managers*, and we shared with you some recommendations to make passwords more difficult to guess.

We talked about the risks of phishing, smishing, vishing, and quishing, and considered how the use of AI tools like *ChatGPT* and *Voice Cloning* makes it easier for cybercriminals to deliver personalized, harder-to-detect scams. We shared with you a list of warning signs to look for in case you receive a suspicious email, text message, or phone call, and listed a few points you should consider before interacting with a QR code. We warned you about the risks of *Clickbait* scams and *Fake News*. We also briefly mentioned the *Missed Call Fraud*, or *Wangiri*, that might have serious financial consequences on the victims.

We spoke about privacy and showed you how to adjust privacy settings in a couple of popular social apps like *Facebook* and *WhatsApp*. We discussed the dangers of social media and dating apps, and analyzed nefarious cybercrimes like *Dating Scams*, *Financial Sextortion*, and *Revenge Porn*. We shared with you some recommendations to avoid falling victims to those crimes, and a couple of tips on which actions to take in case you do.

We discussed the concept of *Digital Footprint* and shared some tips to limit the potentially negative impact it might have on our lives.

We analyzed advantages and disadvantages of

removable media, talked about *USB Drop Attacks*, and listed some best practices to mitigate the security risks they pose. We discussed the risk of *Juice Jacking* when plugging mobile devices into public USB port charging stations, and we gave you a couple of tips on how to protect your equipment when charging is needed.

We warned you about the dangers of public Wi-Fi networks due to the possibility of *Man-in-the-Middle Attacks* with or without *Rogue Access Points*, and we explained the benefits of using a *Virtual Private Network* (*VPN*). We also discussed the risks of pairing wireless devices with Bluetooth and listed some best practices to use this technology more responsibly.

We talked about vulnerabilities, and the importance of promptly patching, turning on or confirming automatic updates, and regularly rebooting electronic devices. We also provided some instructions to manually update apps on Apple and Android smartphones.

Finally, we highlighted the importance of having full, up-to-date, and tested backups stored in an offsite location to be able to face unpredictable events like ransomware attacks, theft, or natural disasters.

Although this book barely scratched the surface of such a complex and ever-evolving discipline as cybersecurity, I am fully aware you may have a lot to digest and reflect on. However, you broke the ice, and now nothing prevents you from developing your knowledge and skills further. Cybersecurity is a hot topic today, and there are plenty of online courses, books, articles, and blogs on the subject.

It is up to you to decide whether to be an easy prey or sell your (digital) life dearly.

I hope you choose the latter.

REFERENCES

Abed A. A. – *Doxware Ransomware: The Insidious Threat to Your Sensitive Information* – https://www.linkedin.com/pulse/doxware-ransomware-insidious-threat-your-sensitive-abed-a-a-/, 2023.

ACSC (Australian Cyber Security Centre) – *How to update your device and software* – https://www.cyber.gov.au/protect-yourself/securing-your-devices/how-update-your-device-and-software, 2024.

Alvarez Technology Group – *Bluetooth Security Issues: Understanding And Preventing Risks* – https://www.alvareztg.com/bluetooth-security-issues-understanding-and-preventing-risks/, 2024.

Aratek – *5 Authentication Factors: A Guide From Passwords to Biometrics* – https://www.aratek.co/news/5-authentication-factors-a-guide-from-passwords-to-biometrics, 2023.

Breachsense – *The 15 biggest data breach examples in history* – https://www.breachsense.com/blog/data-breach-examples/, 2024.

Buxton O. – *Public Wi-Fi: A guide to the risks and how to stay safe* – https://us.norton.com/blog/privacy/public-wifi, 2024.

Carielli S. – *Not So Fast - Mind QR Code Risks, Or Get Ready For Damage Control* – https://www.forrester.com/blogs/not-so-fast-mind-quick-response-qr-code-risks-or-get-ready-for-damage-control/, 2023.

Ceci L. – *Number of sent and received e-mails per day worldwide from 2018 to 2027* – https://www.statista.com/statistics/456500/daily-number-of-e-mails-worldwide/, 2024.

Challita A. – *ZitoVault stopping WannaCry Ransomware* – https://www.linkedin.com/pulse/zitovault-stopping-wannacry-ransomware-antonio-challita/, 2017.

Channel J. – *What is Ransomware?* – https://blog.sucuri.net/2020/02/what-is-ransomware. html, 2020.

Check Point – *What is Locker Ransomware* – https://www.checkpoint.com/cyber-hub/ransomware/what-is-locker-ransomware/, 2024.

Cloudflare – *What was the WannaCry ransomware attack?* – https://www.cloudflare.com/learning/security/ransomware/wannacry-ransomware/, 2024.

Cristello B. – *A Brief History of Cybersecurity* – https://www.linkedin.com/pulse/brief-history-cybersecurity-robert-cristello/, 2023.

Cybercrime Support Network – *Revenge Porn: What It Is and How to Fight Back* – https://fightcybercrime.org/blog/revenge-porn-what-it-is-and-how-to-fight-back/, 2022.

Cybersecurity & Infrastructure Security Agency (CISA) – *Keep Your Device's Operating System and Applications Up to Date* – https://www.cisa.gov/resources-tools/training/keep-your-devices-operating-system-and-applications-date, 2024.

Cybersecurity & Infrastructure Security Agency (CISA) – *Risks of Default Passwords on the Internet* – https://www.cisa.gov/news-events/alerts/2013/06/24/risks-default-passwords-internet, 2016.

Eddy M. – *The Best Password Managers* – https://www.nytimes.com/wirecutter/reviews/best-password-managers/, 2024.

Elegant Themes – *5 Best AI Voice Cloning Tools of 2024* – https://www.youtube.com/watch?v=iEdS_xke-mk, 2024.

ESentire – *2023 Official Cybercrime Report* – https://www.esentire.com/resources/library/2023-official-cybercrime-report, 2024.

exterro – *Largest Password Dump in History Exposes 10 Billion Credentials* – https://www.exterro.com/resources/data-privacy-alerts/largest-password-dump-in-history-exposes-10-billion-credentials, 2024.

Federal Communications Commission (FCC) – *What is 'Juice Jacking' and Tips to Avoid It* – https://www.fcc.gov/juice-jacking-tips-to-avoid-it, 2023.

Fuhrman T. – *Ransomware: Paying Cyber Extortion*

Demands in Cryptocurrency – https://www.marshmclennan.com/assets/insights/publications/2020/november/ransomware-cryptocurrency.pdf, 2020.

Gibson Research Corporation (GRC) – *How Big is Your Haystack? ...and how well hidden is your needle?* – https://www.grc.com/haystack.htm, 2012.

Glassner A. – *Cyber Civil Rights Initiative Combats Revenge Porn* – https://cybersecurityventures.com/cyber-civil-rights-initiative-combats-revenge-porn/, 2021.

Global Information Assurance Certification (GIAC) – *The Default Password Threat* – https://www.giac.org/paper/gsec/317/default-password-threat/100889, 2005.

Griffith E. – *Stop Changing Your (Strong, Unique) Passwords So Much* – https://uk.pcmag.com/password-managers/135788/stop-changing-your-strong-unique-passwords-so-much, 2021.

Griffiths C. – *The Latest 2024 Ransomware Statistics* – https://aag-it.com/the-latest-ransomware-statistics/, 2024.

Guarino B. – *Defend Yourself against AI Impostor Scams with a Safe Word* – https://www.scientificamerican.com/article/a-safe-word-can-protect-against-ai-impostor-scams/, 2024.

Guha A. – *Understanding Wangiri Scams: Impact, Mechanics, and Protection Strategies* – https://www.

linkedin.com/pulse/understanding-wangiri-scams-impact-mechanics-protection-abhirup-guha-lrxoc/, 2024.

Haan K. – *The Real Risks Of Public Wi-Fi: Key Statistics And Usage Data* – https://www.forbes.com/advisor/business/public-wifi-risks/, 2023.

Hill S. – *How to Reboot Your Gadgets and How Often to Do It* – https://www.wired.com/story/how-to-reboot-your-gadgets/, 2022.

IBM – *What is a threat actor?* – https://www.ibm.com/topics/threat-actor, 2024.

Ibrahim M. – *10 Benefits of Multi-Factor Authentication (MFA)* – https://supertokens.com/blog/benefits-of-multi-factor-authentication, 2024.

Insight IT Team – *The Hidden Cybersecurity Risks of Printers: What You Need to Know* – https://www.insightit.com.au/the-hidden-cybersecurity-risks-of-printers/, 2024.

Jogi P. – *What is a QR Code? Its Usage, Vulnerability, Advantages, and Comeback Story* – https://www.ssl2buy.com/cybersecurity/qr-code-usage-vulnerability-advantages, 2023a.

Jogi P. – *Understanding QR Code Risks, Scams, Examples & Best Security Practices* – https://www.ssl2buy.com/cybersecurity/qr-code-risks-scams-examples-security-practices, 2023b.

Keepnet – *2024 QR Code Phishing Trends: In-Depth Analysis of Rising Quishing Statistic*s – https://keepnetlabs. com/blog/2024-qr-code-phishing-trends-in-depth-analysis-of-rising-quishing-statistics, 2024.

King T. – *45 World Backup Day Quotes from 32 Experts for 2023* – https://solutionsreview.com/backup-disaster-recovery/world-backup-day-quotes/, 2023.

Landesman M. – *How to Disable AutoRun and AutoPlay for External Devices* – https://www.lifewire.com/disable-autorun-on-a-pc-153344, 2022.

Larson S. – *A smart fish tank left a casino vulnerable to hackers* – https://money.cnn.com/2017/07/19/technology/fish-tank-hack-darktrace/index.html, 2017.

Martin A. – *UK becomes first country to ban default bad passwords on IoT devices* – https://therecord.media/united-kingdom-bans-defalt-passwords-iot-devices, 2024.

Masjedi Y. – *How To Spot a Scammer on a Dating Site: 9 Warning Signs* – https://www.aura.com/ learn/how-to-spot-a-scammer-on-a-dating-site, 2023.

Maslin Nir S. – *A flirty 'Chelsea' asked for nudes. It was a trap.* – The New York Times, May 18-19, 2024.

Meeuwisse R. – *Cybersecurity for Beginners* – CyberSimplicity, 2015.

Microsoft – *Create and use strong passwords* –

https://support.microsoft.com/en-us/windows/create-and-use-strong-passwords-c5cebb49-8c53-4f5e-2bc4-fe357ca048eb, 2024.

Milgram S. – *The small world problem* – Psychology Today, 1967.

Morgan Stanley – *Strategies to Help Protect Your Digital Footprint* – https://www.morganstanley.com/articles/digital-footprint-protection-strategies, 2023.

Morgan Stanley – *The Importance of Data Backups* – https://www.morganstanley.com/articles/data-backup-importance-cybersecurity, 2024.

National Cyber Security Centre (NCSC) – *Top tips for staying secure online* – https://www.ncsc.gov.uk/collection/top-tips-for-staying-secure-online, 2021.

National Institute of Standards and Technology (NIST) – *NIST Special Publication 800-63B, Digital Identity Guidelines* – https://pages.nist.gov/800-63-3/sp800-63b.html, 2017.

National Institute of Standards and Technology (NIST) – *Vulnerability Metrics* – https://nvd.nist.gov/vuln-metrics, 2024.

Naval Criminal Investigative Service (NCIS) – *NCIS Warns Sailors following Spike in 'Sextortion' Cases* – https://www.navy.mil/Press-Office/Press-Releases/display-pressreleases/Article/2257031/ncis-warns-sailors-following-spike-in-sextortion-cases/, 2016.

Northwestern University – *The Benefits of Restarting Your Computer Regularly* – https://services.northwestern.edu/TDClient/30/Portal/KB/ArticleDet?ID=2584, 2024.

Paulyn M. – *What Is the Meaning of Clickbait and Is It Dangerous?* – https://www.avg.com/en/signal/what-is-clickbait-is-it-dangerous, 2024.

Petrosyan A. – *Annual number of ransomware attempts worldwide from 2017 to 2023* – https://www.statista.com/statistics/494947/ransomware-attempts-per-year-worldwide/, 2024.

Petrosyan A. – *Number of internet and social media users worldwide as of July 2024* – https://www.statista.com/statistics/617136/digital-population-worldwide/, 2024.

Puri N. – *Is That QR Code Safe? How To Evaluate and Stay Protected* – https://www.uniqode.com/blog/qr-code-security/how-to-check-if-a-qr-code-is-safe, 2024.

Silkalns A. – *Password Statistics (How Many Passwords Does An Average Person Have?)* – https://colorlib.com/wp/password-statistics/, 2024.

Solá A. T. – *Romance scams cost consumers $1.14 billion last year. It's a 'more insidious' fraud, expert says* – https://www.cnbc.com/2024/07/03/heres-how-to-avoid-romance-scams-which-cost-consumers-1point14-billion-last-year.html, 2024.

St. John M. – *Cybersecurity Stats: Facts And Figures You Should Know* – https://www.forbes.com/advisor/education/it-and-tech/cybersecurity-statistics/, 2024.

Stickley J. – *Jim Stickley demonstrates the risks of QR codes with Jeff Rossen* – https://www.youtube.com/watch?v=YQlhHXz27MI&t=106s, 2023.

Stouffer C. – *Bluetooth security risks to know (and how to avoid them)* – https://us.norton.com/blog/mobile/bluetooth-security, 2022.

Texas Woman's University (TWU) – *Internet Dating and Romance Scams* – https://twu.edu/technology/information-security/cyber-hygiene-and-guidance/internet-dating-and-romance-scams/, 2024.

Touchtidou S. – *Artificial intelligence fuelling global surge in cybercrime* – https://www.euronews.com/2024/05/08/cybercrime-on-the-rise-thanks-to-artificial-intelligence, 2024.

Trevino A. – *What is Juice Jacking?* – https://www.keepersecurity.com/blog/2023/07/10/what-is-juice-jacking/, 2023.

Trevino A. – *Why is Password Security Important?* – https://www.keepersecurity.com/blog/2022/09/14/why-is-password-security-important/, 2022.

Vailshery L. S. – *Number of Internet of Things (IoT) connections worldwide from 2022 to 2023, with forecasts from 2024 to 2033* – https://www.statista.com/statistics/1183457/iot-connected-devices-worldwide/, 2024.

Verizon – *Best practices for using removable media devices* – https://www.verizon.com/about/blog/using-removable-media-devices, 2024.

Vigderman A. – *The Best VPN Services of 2024* – https://www.security.org/vpn/best/, 2024.

Violino B. – *AI tools such as ChatGPT are generating a mammoth increase in malicious phishing emails* – https://www.cnbc.com/2023/11/28/ai-like-chatgpt-is-creating-huge-increase-in-malicious-phishing-email.html, 2023.

Winder D. – *Honda Hacked: Japanese Car Giant Confirms Cyber Attack On Global Operations* – https://www.forbes.com/sites/daveywinder/2020/06/10/honda-hacked-japanese-car-giant-confirms-cyber-attack-on-global-operations-snake-ransomware/, 2020.

World Economic Forum – *Charted: There are more mobile phones than people in the world* – https://www.weforum.org/agenda/2023/04/charted-there-are-more-phones-than-people-in-the-world/, 2023.

Yasar K. – *Man-in-the-middle attack (MitM)* – https://www.techtarget.com/iotagenda/definition/man-in-the-middle-attack-MitM, 2022.

www.ingramcontent.com/pod-product-compliance
Lightning Source LLC
Chambersburg PA
CBHW070934210326
41520CB00021B/6940